Milton Whitney

The Alkali Soils of the Yellowstone Valley

Milton Whitney

The Alkali Soils of the Yellowstone Valley

ISBN/EAN: 9783743376991

Manufactured in Europe, USA, Canada, Australia, Japa

Cover: Foto ©berggeist007 / pixelio.de

Manufactured and distributed by brebook publishing software
(www.brebook.com)

Milton Whitney

The Alkali Soils of the Yellowstone Valley

LETTER OF TRANSMITTAL.

U. S. DEPARTMENT OF AGRICULTURE,
DIVISION OF SOILS,
Washington, D. C., October 22, 1898.

SIR: I have the honor to transmit herewith the manuscript for a bulletin on the results of some investigations of the alkali soils of the Yellowstone Valley. The manuscript is very fully illustrated in order that the practical side of the work can be plainly shown. Plates I–V, XVI. XVII, are from photographs taken by A. B. Rumsey, Billings, Mont. Plate XV is from a photograph by Dr. F. W. Traphagen of Bozeman, Mont.

I recommend that this be published as Bulletin No. 14 of this division.

Respectfully,

MILTON WHITNEY,
Chief of Division.

Hon. JAMES WILSON.
Secretary of Agriculture.

2

PREFACE.

In August, 1897, I had a conference with the general manager and the land commissioner of the Northern Pacific Railroad at St. Paul, Minn., in regard to the character of the land through which their road runs and through which I expected to pass in a trip of reconnoissance through the Northwest. They both urged me to stop at Miles City and Billings, Mont., in the Yellowstone Valley, to investigate the alkali lands. They stated that serious trouble was threatened at these places from the rise of alkali in the soils. In opening up the country there had been no suggestion of trouble with alkali, and it was only beginning to assume alarming proportions now after some ten or twelve years of irrigation.

I stopped at Billings, and was driven over a large area of the valley at that place by several gentlemen interested in the matter. The conditions seemed so very important and there was so much uncertainty as to exactly what the conditions were under the surface that I sent back to Washington for an equipment for the electrical determination of the salt content of the soils, and returned to Billings when notified that this had arrived. The brief examination that was possible in the short time at my disposal satisfied me that further investigations would give results of very great practical value by showing the farmers just what the conditions were in the soil; what they had to fear if existing methods were continued; what they had to guard against to prevent the spread of alkali, and what they would have to contend with in reclaiming some of the already abandoned lands.

The following spring Mr. Thomas H. Means, an assistant in this division, was sent out to Billings to make a more detailed investigation. Owing to the small amount of money available he was only able to stay there about six weeks, but he was charged to collect data as to the general distribution of the salts with reference to the topography of the land and the position of the irrigating ditch, and to make such a detailed examination of a small area that an underground map could be constructed, showing the amount and distribution of the alkali salts under the surface. This work has been very thoroughly done, as will appear in the following pages.

The results of this investigation appear of so much practical value that arrangements are being made to continue the work and to make an underground survey of the entire valley. It is estimated that in six months' time and at an expense not exceeding $1,200, a sum utterly insignificant in proportion to the great value of the probable results of such an expedition, the necessary data could be collected from Billings eastward to Miles City or possibly Glendive and suitable underground maps prepared, with a view to preventing such serious calamities as have overtaken some of these lands in opening up new districts and for the reclamation of abandoned irrigated lands in the Yellowstone Valley. I sincerely trust that arrangements may be consummated for the completion of this survey.

It must not appear to those unacquainted with the subject that the rise of alkali and the disastrous effects following the application of irrigation waters is peculiar to the Yellowstone Valley. These are problems which have to be confronted in all arid regions the world over and wherever irrigation is practiced. The Yellowstone Valley was really selected to start the alkali work of the division because there the problem is so simple and the conditions can be so easily controlled. All of the irrigation districts of the country contain more or less alkali and are subject to the evil effects of overirrigation, and the community is to be congratulated where the conditions are so simple and so easily controlled as in the Yellowstone Valley.

M. W.

CONTENTS.

ILLUSTRATIONS.

TEXT FIGURES.

PLATES.

THE ALKALI SOILS OF THE YELLOWSTONE VALLEY.

INTRODUCTION.

The Yellowstone Valley in Montana is approximately 400 miles in length, but the area which has been especially considered in these investigations is that part of the valley between Billings and Glendive, a distance of about 250 miles. All of the detailed work, however, mentioned in this report was done in the immediate vicinity of Billings. The valley at this point is about 6 miles in width. Irrigation has been practiced for twelve or fifteen years. The water for the main ditch supplying the valley at this point is taken out of the river nearly 40 miles above the town of Billings.

When the country was first settled and, indeed, above the ditch at the present time, the depth to standing water in the wells was from 20 to 50 feet, and there were no signs of alkali on the surface of the ground. Under the common practice of irrigation, however, an excessive amount of water has been applied to the land, and seepage waters have accumulated to such a degree that water is now secured in wells at a depth of from 3 to 10 feet in the irrigated district, while many once fertile tracts on the lower levels are already flooded, and alkali has accumulated on them to such an extent that they are mere bogs and swamps and alkali flats, and the once fertile lands are thrown out as ruined and abandoned tracts. This injury, while not very widespread as yet, has been so serious where the results have appeared that the owners of the land are naturally very much concerned. Fortunately they are prepared to receive kindly and to take advantage of all the information that can be thrown upon the condition.

Many theories have been advanced by the landowners as to the source of the alkali and as to the conditions under the surface. The belief is widespread that the alkali flats could probably be reclaimed by flooding the surface during a dry season and washing the crust off. Our preliminary investigations showed them plainly that the crust contained only a small proportion of the alkali, and convinced them that this method would not be efficient. They are beginning to realize, however, that the less water they use the safer; but still many of the planters distrust this suggestion as a possible device of the ditch management to restrict the use of water in order that it may go further and supply a larger number of customers.

There is generally little system in the application of water to the land. Very few of the planters know how much they use, and none of them pretend to know how much they need. The water is applied

7

when the surface appears dry, and it is then applied in such excess that much of it can not evaporate and can not seep down through the subsoil before the surface again dries out and a further heavy application of water is applied by flooding. The planters do not realize (indeed, how can they, not seeing the conditions under the surface) that the water is accumulating there, and although it may not be rising rapidly enough to injure themselves, it may be seeping down and gradually inundating their unfortunate neighbors on lower levels. In every case of injury to crops which was examined around Billings the first trouble was shown to be an accumulation of seepage water near the surface of the ground. This water is not at first excessively alkaline, but from the continual evaporation of so much water from the surface the soluble salts accumulate to an excessive and ruinous extent, prejudicial to the growth of all cultivated plants.

Hilgard has been pointing out for years that the only safe practice in bringing a new area under the ditch, in a soil which is at all likely to have alkali, is to use water very sparingly and to keep the surface under very thorough cultivation so that a minimum amount of water shall evaporate from the surface of the ground and that there shall be no accumulation of seepage waters in the subsoil.

If these investigations do no more than to show the full results of overirrigation and the necessity of intelligent and careful application of water to the soil and the importance of underdrainage, where the alkali can not be otherwise controlled or removed, the authors will be well repaid.

Proper irrigation in an arid region furnishes an ideal condition of crop production. In practice, however, the method of applying water to the land is extremely crude, and there is really little cause to wonder that much harm has been done through overirrigation, and it is no wonder that very serious trouble has occurred through the rise of soluble salts. There is less excuse for this wasteful use of water now, however. In the first place water is more valuable, and if three areas can be supplied with sufficient water from what is now applied on one area it would result not only in the advantage of an increased area brought under the ditch, but in the distinct advantage of a smaller application per acre. Furthermore, we understand now that different plants require different amounts of water for their best development, and even that the same crop requires varying amounts of water during the various periods of its development. Different soils require different amounts of water, depending upon their texture.

There is reason to hope that these principles will come to be recognized and controlled in the field as they are in the commercial greenhouses of the present time. Furthermore, the Division of Soils has invented (described heretofore) a very ingenious instrument for determining the moisture content of soils through their electrical resistance. With this instrument the amount and fluctuations of the water content near the surface of the ground or at any desired depth can be readily

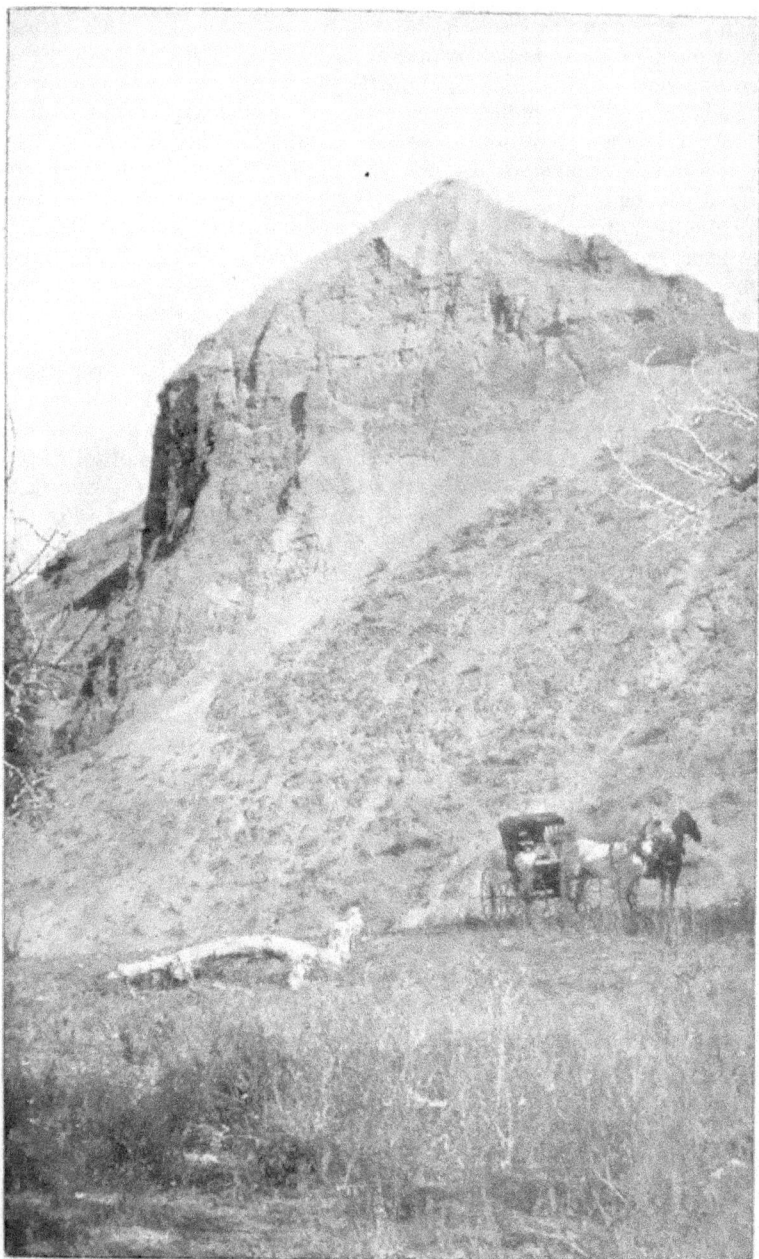

A NEAR VIEW OF THE SLATE BLUFF SOUTH OF THE RIVER, WITH THE TALUS SLOPE
FORMED FROM THE DISINTEGRATION AND BREAKING DOWN OF THE ROCKS.
These rocks are filled with veins of gypsum and alkali salts

determined, and by the use of such instrument it is quite possible, with an irrigation plant, to maintain any desired water content near the surface, and to prevent an undue accumulation in the subsoil. So important is this considered that a very simple and cheap modification of the instrument has been devised for use in irrigated fields, to cost probably not over $10 apiece, so that for a moderate expense enough of these could be distributed, even over a large plantation, so that the fluctuations of the water content of the soils of the different fields could be closely studied and controlled.

ORIGIN OF THE ALKALI SALTS.

Any excessive accumulation of soluble mineral salts in the soil is popularly spoken of as "alkali" in the West. The term, therefore, as popularly used and as used in this bulletin, does not necessarily refer to material of an alkaline or basic nature. The alkali soils of the West are of two principal classes: The alkaline carbonates, or black alkali, usually sodium carbonate, is the worst form of alkali, actually dissolving the organic materials of the soil and corroding and killing the germinating seed or roots of plants; the white alkalies, the most common of which are sodium sulphate, sodium chloride, magnesium sulphate, magnesium chloride, and occasionally, as in northern Nevada, some of the borates, are not in themselves poisonous to plants, nor do they attack the substance of the plant roots, but are injurious when, owing to their presence in excessive amounts, they prevent the plants from taking up their needed food and water supply.

Field work in the alkali soils at Billings has shown that when the concentration of the salts in active solution in the soil moisture is as high as 1 per cent the limit of most cultivated plants is reached. Further concentration kills all our ordinary agricultural plants. This is at once understood when the osmotic properties of the salts in root-cell solutions are considered. Dyer found that the acidity of root saps of a number of our common agricultural plants was equivalent in concentration to a 1 per cent solution of citric acid. The osmotic pressure in 1 per cent citric acid solution is roughly the same osmotic pressure of the 1 per cent solution of the soluble matter in the alkali soil at Billings, Mont.; so if the concentration of the soil solutions becomes greater than a 1 per cent solution the osmotic pressure of the solution outside of the cell is greater than the pressure of the solution inside the cell and the cell is unable to absorb water, the plant turns yellow, just as it would do if the soil contained too little moisture, and finally dies. This is but a rough comparison for illustration purposes only, for it is not to be supposed that the soluble matters of the root cell are all acids and acid salts. There are, unquestionably, neutral salts present which influence osmotic pressure. The real osmotic pressure in the root cells is so difficult a matter to determine that relative value only can thus be very roughly approximated for the osmotic pressure inside the cell and in the soil moisture in the alkali lands at Billings.

Hilgard states that plants will stand a larger percentage of sodium sulphate than of sodium chloride. The reason is that for the same concentration, expressed in per cents, the osmotic pressure of the sodium chloride is greater than the osmotic pressure of the sodium sulphate. Consequently a plant is able to bear more of the sulphate than of the chloride.

Some of our agricultural plants show a greater ability to stand salts in solution than others. The reason for this is probably explained by the fact that in some plants the concentration of the cell solution is greater than in other plants. The osmotic pressure of the sugar-beet root cells, for example, may be greater than those of other plants, and for that reason the beet can grow when the soil solutions are of higher osmotic pressure.

It has been found that the solid grains of soil have the remarkable power of absorbing or concentrating a portion of the salts on their surface and thus withdrawing them from active solution. This is of the greatest practical importance, as otherwise the soil moisture would quickly become saturated with salts and rendered totally unfit for agricultural plants. As a matter of fact, in consequence of this condensing power, in no case was the concentration of the soil moisture found to exceed 3 per cent, although the salts were quite soluble and were crystallized out on the surface of the ground.

The amount of soluble salts which plants can stand depends upon the character of the salt, the character of the soil, and the kind of plant. Hilgard states that few plants can bear as much as 0.1 of 1 per cent of sodium carbonate, or about 3,500 pounds per acre to a depth of 1 foot; of sodium chloride, about 0.25 per cent, and of sodium sulphate most plants can grow with 0.45 to 0.50 per cent present, and are affected by even less salts in the sandy lands than on heavy clay or gumbo lands.

The soluble salts at Billings, Mont., are mainly sodium sulphate and magnesium sulphate, with none of the sodium carbonate and with but a trace of sodium chloride. The soils also contain a considerable amount of the difficultly soluble calcium sulphate or gypsum.

The following table gives the mean of five analyses of the soluble salts by Dr. F. W. Traphagen, of the Montana Station. From his researches the composition of the salts appears to be very constant throughout the valley.

Composition of the soluble salts at Billings, Mont.

	Per cent.
Sodium sulphate	57.44
Magnesium sulphate	27.59
Calcium sulphate	13.05
Potash sulphate	1.55
Silica	.36
	99.99

The sodium sulphate and the magnesium sulphate accumulate to such an extent in the draws and low levels that all agricultural crops

PLATE III.

A NEAR VIEW OF THE SANDSTONE BLUFF ON NORTH SIDE OF VALLEY.

are destroyed and the land abandoned. The important question at once arises as to the origin of these salts, what they have been derived from, how they were distributed in the soils before the practice of irrigation had been introduced, and how they leach and shift about in the soil as a result of overirrigation.

It is so important that a clear conception be obtained of this subject that a short space will be given to a consideration of the formation of soils and to the cause of the accumulation of such large amounts of soluble salts in the soils of the arid regions.

THE FORMATION OF SOILS.

Soils are derived from the disintegration and decomposition of rocks. It is unnecessary to speak here of the influences which occasion this decay, but it is sufficient to say that the process of the breaking down of rocks and the formation of soils is taking place with varying rapidity at every point where rocks are exposed to the action of the weathering influences.

During the weathering process the rock crumbles and breaks down into the minute particles composing the resultant soil. While the rock itself had been hard and continuous, with little appreciable space for the absorption of water and air, the resultant soil is permeated with air spaces aggregating about 50 per cent of the bulk of the soil. As one of us pointed out years ago, in the formation of the soil at least half of the bulk of the rock has been leached out and carried away, or else the material of which the rock was composed has swollen to twice the superficial volume it originally occupied. Chemical investigation has shown, however, that a very large percentage of the original rock material is actually dissolved and carried off.

The following table, compiled from the writings of G. P. Merrill, will give a general idea of the magnitude of the changes due to the solubility of the rock in the formation of soils and will give an idea of the enormous amount of material that must be decomposed and removed in order to form soil one foot in depth on an acre of land:

Amount of soluble matter removed in the decomposition of rocks and the formation of soils.

Kind of rock.	Locality.	Rock removed by solution for each acre-foot of soil formed.
		Per cent. Tons.

During the weathering process, therefore, and especially in the slow decomposition of the minerals of the rock or soil, certain portions, and in the aggregate very large portions, of the rock are rendered soluble and accumulate within the soil as soluble salts, or are carried off in solution and washed off into the ocean or some inclosed basin which receives the drainage waters of the district. Through the action of water alone, or through the action of water containing minute quantities of carbonic or other organic acids, rocks of all kinds are thus slowly dissolved. In the humid portions of the United States approximately 50 per cent of the rainfall seeps down through the subsoil and runs off through the rivers into the ocean. This constant seepage through the soil prevents any excessive accumulation of soluble matters within the soil. In the arid portions of the country, however, under scanty rainfall there is not sufficient underdrainage or seepage waters to carry off the amount of salts rendered soluble in the decay of rocks, and we have, therefore, as a rule, a much larger soluble salt content in soils of the arid regions than in the soils of the humid regions. The greatest contrast is of course shown in the case of the very soluble salts of sodium, magnesium, calcium, and potassium. These are very completely removed from the soils of the humid regions. The very fact of the easy solubility of the salts renders them dangerous in the practice of agriculture when they accumulate, as in the soils of the arid West.

The kind of salt depends, of course, upon the kind of rock, and to a certain extent upon the factors producing decomposition.

The following table gives a list of the alkali-bearing minerals occurring in primary rocks as the ultimate source of the alkali of soils:

Feldspars:	Per cent of alkalies	Micas:	Per cent of alkalies
Orthoclase	17	Muscovite	12
Microcline	17	Biotite	10
Albite	12	Phlogopite	9
Oligoclase	9	Nepheline	24
Andesite	8	Leucite	21.5
Labradorite	4	Sodalite	26
Bytownite	3.5	Hauyn	17
Anorthite	2		

Some of these alkali-bearing minerals are very generally present in the primary rocks from which the soils have all ultimately been derived, but they are of course usually mixed with other minerals, so that the total percentage of alkalies in the rock are not so great as would appear from these minerals.

In the formation of sedimentary rocks these soluble salts are often deposited in local areas to such an extent as to form a marked feature of the subsequent geological formation of the land. Where the sedimentary rocks are formed in inclosed basins the soluble salts which have leached out of the soils in the decomposition of the rocks over the drainage area are concentrated by the evaporation of the waters of the inland lake or sea and the salts are precipitated out of solution and deposited in layers, either in a pure state or mixed with the sedimentary material that is brought down with them. Such a concentration and precipitation of salts is seen in the Great Salt Lake district of Utah, in the Dead Sea, and in the many salt lakes of central Asia. To a modified and limited extent the same accumulation goes on in the salt marshes along the coast.

The water of the ocean at the present time contains about 3½ per cent of soluble salts, having the following composition:

Composition of the dissolved salts of sea water.

	Per cent
Sodium chloride	78
Magnesium chloride	11
Magnesium sulphate	5
Calcium sulphate (gypsum)	4
Potassium sulphate	2

Of these salts the calcium sulphate or gypsum is very slightly soluble, and on concentration of the solution this is the first salt to be deposited. The soils of the arid West are particularly rich in gypsum and carbonate of lime, and there are many extensive deposits of gypsum in beds, which are capable of being worked commercially. Such depositions from the evaporation of vast bodies of water are believed to be the origin of extensive deposits of gypsum. If the concentration continues sufficiently long the sodium chloride and the sulphates are deposited. In the salt mines at Stassfurt, Germany, the rock salt, which was formerly the substance of principal value to be mined, was overlaid with sulphates of soda, magnesia, and potash, and chlorides of potash and magnesia, while the rock salt was underlaid with very deep deposits of calcium sulphate or gypsum. There are many places in the West where the salts have deposited and where they are now depositing, mixed with large quantities of sedimentary materials and not in pure deposits of salts. In later geological ages, when these sedimentary rocks are raised to the surface and again subjected to erosion, the salts are again exposed to the solvent action of meteoric waters.

In humid regions the salts are removed by leaching almost as rapidly as the disintegration of the rocks proceeds, and there is little evidence in the soil of their former existence. However, in wells driven to the lower depths of the subsoil, where the movement of the water has been slow and the soluble salts have been only slowly and partially removed, the water is frequently charged with the soluble mineral salts. Deep-seated springs also very frequently give evidence of soluble mineral matters within the depths of the rocks, as seen in the many mineral springs in all sections of the country.

Where the rainfall is slight, as in the arid regions, and the leaching of the salt and the underlying rocks is small, the removal of the soluble matter has not gone on to such a degree. Immediately below the surface there are evidences of their presence, and in localities of very small rainfall and very scant drainage, and where the evaporation of the soil moisture has been excessive, the salts often appear as efflorescence on the surface of the ground.

The nature of the rock gives some indications of the relative amount of alkali to be found in it, as a greater amount of salts is deposited in the slow-forming shales than in the more rapid formation of the sandstone rocks. The source of the alkali salts at Billings is in the shales forming the bluffs on the south side of the valley, rather than in the sandstone of the north bluff.

The mineral elements occurring as "alkalies" in the arid regions the world over consist mainly of sodium, magnesium, and potassium, combined as carbonates, chlorides, sulphates, and occasionally as nitrates, phosphates, and borates. The alkalies themselves are derived originally from the decomposition of the feldspars and micas and three or four other minerals mentioned in the list on page 12.

The origin of the carbonic acid of the carbonates is either the atmosphere or decaying vegetation. The water percolating through the soil is charged more or less with carbon dioxide and the solubility of the rocks is very much greater in these carbonated waters than in pure water. The most important factor in rock decomposition, especially in deep-seated rocks, is the hydration—that is, the chemical combination of water with the minerals forming the rock. This action finally results in the liberation of the alkalies as carbonates and the formation of free silica and the simple silicates of alumina, magnesia, etc. The complete kaolinization of one of the feldspars indicates the processes of the formation of alkaline carbonates in the decomposition of alkali-bearing minerals.

The following equation represents the full and complete decomposition of feldspar and the formation of kaolinite and free silicic acid and potassium carbonate:

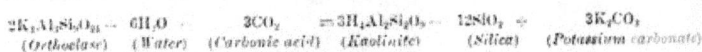

$$2K_2Al_2Si_6O_{16} - 6H_2O \quad 3CO_2 = 3H_4Al_2Si_2O_9 - 12SiO_2 + 3K_2CO_3$$
(Orthoclase) (Water) (Carbonic acid) (Kaolinite) (Silica) (Potassium carbonate)

There are, of course, many stages before the complete decomposition of the feldspar, and more complex reactions probably occur during the process, but the final products are indicated in the general changes noted by the equation, and this is the prime source of the alkaline carbonates.

The source of the sulphuric acid, which is in combination in the sodium sulphate of our alkaline plains and in combination in calcium sulphate or gypsum, is probably the oxidation of free sulphur in and around volcanic districts in the alteration of pyrite and marcasite, and in the subsequent deposition of sulphates from sea water.

Maxwell, in writing upon the Hawaiian lavas and soils, notes the occurrence of steam and water vapor containing 5 per cent of free sulphuric acid. Such acid waters coming in contact with the rocks dissolve the lime, magnesia, alumina, iron, and the alkalies with the formation of gypsum, magnesium sulphate, iron sulphates, alkali sulphates, and

several alums. The sulphates of iron thus formed readily decompose in the presence of water, forming oxides of iron, other forms of sulphate, and even free sulphuric acid. The sulphates and free sulphuric acid in contact with lime or alkali carbonates react with the formation of lime or alkali sulphates and iron carbonates or oxides, so that in the presence of pyrites or marcasite the lime or alkaline carbonates will be changed, to a certain extent at any rate, into sulphates.

There are other reactions equally important to the agriculturist besides those which have just been noted.

Hilgard has repeatedly called attention to the reactions between lime salts and the salts of the alkalies.

Where sodium or potassium carbonates or chlorides are associated with calcium sulphate in a well-aerated soil, a reaction takes place in which sodium or potassium sulphate and calcium chloride or carbonate are formed. The calcium chloride is extremely soluble and easily leached from the soil if there is any chance at all of its being carried off by drainage waters. The calcium carbonate is difficultly soluble and would remain in the soil as limestone. On the other hand, where sodium or potassium sulphate exists in the soil, together with lime carbonate and in the presence of an excess of carbonic acid or in the presence of supercarbonates of the alkalies, the reverse action takes place and carbonate of soda or potash is formed, together with the sulphate of lime.

Hilgard shows this reaction actually taking place in certain conditions of the soil. He points out that the carbonate of soda and sulphate of lime may occur in the bottom of a slight depression, where the soil is moist, while the sodium sulphate and calcium carbonate will be found around the edges where the soil is better drained. As a practical application of this matter, he urges the use of gypsum or sulphate of lime in the reclaiming of lands containing the black alkali or carbonate of soda, and at the same time points out the necessity of thorough drainage in connection with the application of gypsum; otherwise the application will do no good at all.

These various reactions and properties of the so-called alkali salts indicate the methods for the reclamation of the alkali lands. In the case of the carbonates the course recommended by Hilgard is unquestionably the proper one—to treat the soil with heavy applications of gypsum and insure thorough drainage, so as to have the soil well aerated. In the case of an excess of sodium chloride, which is very soluble and easily leached out of a soil, it is only necessary to flood the soil and remove the excess of salt in this way. It is essential, however, that the soil so treated should have good underdrainage in order that the water applied at the surface may percolate through and actually carry off the excessive amount of the soluble sodium chloride. No application of any kind will be beneficial, as the sodium chloride is as simple a salt as one can have and quite harmless, except when present in extraordinary amounts.

The treatment of soils containing sodium sulphate is more difficult than in the case of the chloride, as the salt is less easily leached from the soil. Here again no application, however, can be made, the sulphate, like the chloride, being injurious only when in large excess. It will be shown further on that the surest plan in the cultivation of these alkali soils is to use care in applying the water, so that there shall be no accumulation of the salts at the surface, and, as Hilgard has repeatedly recommended, the cultivation should be very thorough so as to prevent, so far as possible, the evaporation of the water from the surface of the ground. When the salts have once accumulated, however, there is nothing to do but wait for them to gradually leach away through the drainage and seepage waters or to thoroughly underdrain the land with tile drains, and so hasten the reclamation.

THE GEOLOGICAL STRUCTURE OF THE VALLEY AT BILLINGS.

The Yellowstone River at Billings traverses a broad valley from 6 to 10 miles wide. A sketch map (fig. 1) is given of a portion of the eastern part of the valley, showing the relative positions of the irrigating ditch, the river, the site of Billings, and the bluffs on either side. The valley is bordered on either side with high bluffs. On the north side the bluff is of sandstone. On the south are steep ragged hills of blue shale. The shale or slate dips under the sandstone and is found at various depths throughout the valley.

Fig. 2 gives a diagrammatic illustration of the relative positions of these two classes of rocks. The section is drawn from north to south across the valley.

The sandstone is a gray siliceous stone, rising abruptly to a height of from 200 to 500 feet above the general level of the valley. The rocks are of the Upper Cretaceous age included in the great series of cretaceous rocks extending over the greater part of eastern Montana. The sandstone forms an excellent building stone if protected from the alkaline waters of the locality, but under the influence of these salts the stone breaks down and crumbles into loose sand. The sand grains in the rock are about 0.5 to 0.1 millimeter in diameter, and water readily penetrates the pores of the stone. Small but perceptible quantities of magnesium and sodium sulphates are to be found throughout the sandstone rocks, and where evaporation has gone on from the surface for a considerable time white crusts of these salts form on the surface.

Water seems to percolate through fine veins and cracks in the rocks and issue at the sides and the foot of the bluffs, in many places giving rise to springs of alkali water containing a greater or less amount of the soluble sulphates. Wherever this water issues, the rock disintegration has gone on to a great extent, and grottoes are formed in this way in the bluff. There are frequently hard compact layers throughout the rock, and the soluble salts accumulate just above these and are seen on the surface in the canyons and exposed bluffs as white efflorescence or layers lying parallel to the layers of hard and impervious material.

GENERAL VIEW OF THE VALLEY LOOKING TOWARD THE TOWN OF BILLINGS, SHOWING THE CHARACTER OF THE BOTTOM LANDS AND THE DRAINAGE FROM

17

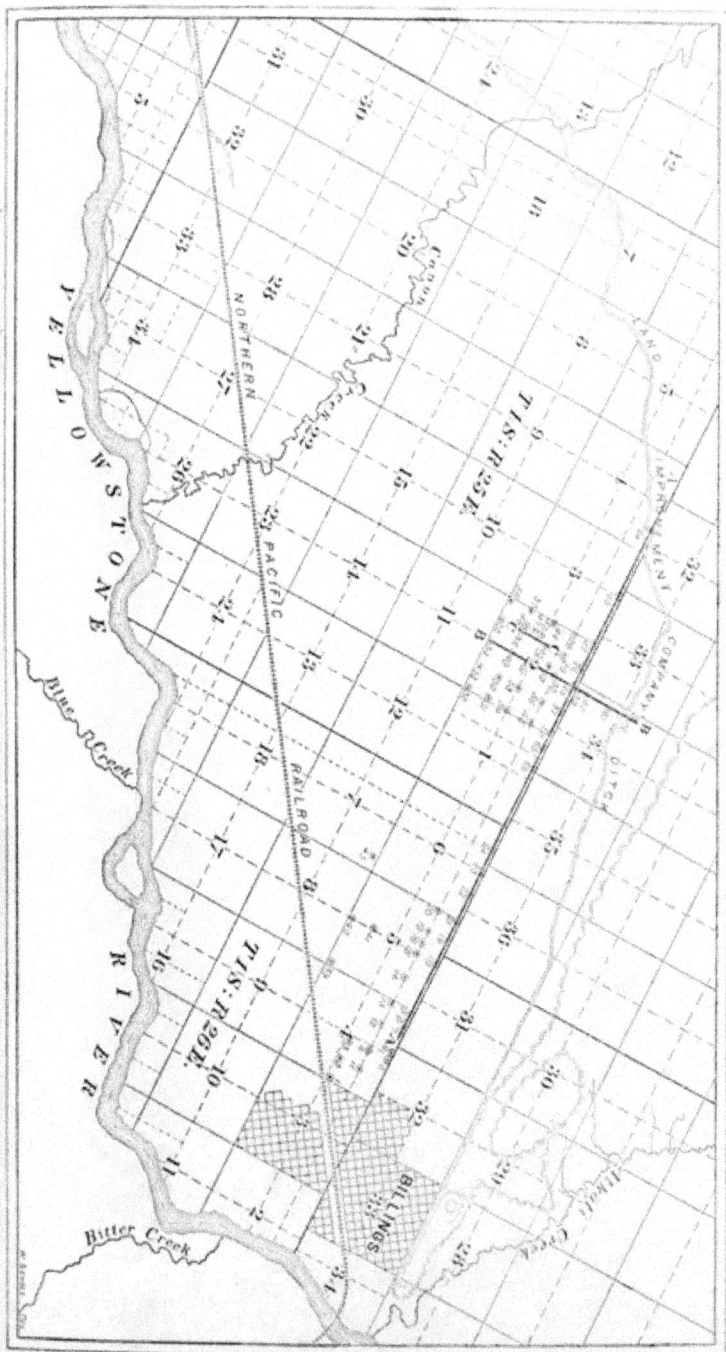

FIG. 1.—Sketch map of a portion of the Yellowstone Valley near Billings, Mont., showing the area examined and the location of the borings.

7769—No. 14——2

The main body of the sandstone has been eroded rather unevenly and the peculiar pinnacle-shaped structure is seen that is so characteristic of the Bad Lands. The face of the bluff, however, is completely rounded by weathering, and the general ragged character is smoothed off by the continual falling of the débris, due to the rapid disintegration of the rock when freely exposed to the weathering influences.

Underlying the sandstone and coming out from under the sandstone bluff there is a fine blue shale or slate which extends to an unknown depth. In an attempt to get artesian water at Billings a well was driven 900 feet through this shale. No deeper record than this has ever been made at this place. The shale rises up from beneath the sandstone and forms the rough angular blue hills on the south side of the valley.

The shale is penetrated with numerous fine cracks and joints running in all directions, and these are filled with fibrous gypsum. Many cavities also are found filled with gypsum and calcium carbonate. Everywhere throughout the shale large quantities of sodium and magnesium sulphates are found, which appear as white efflorescence where evaporation has taken place. These blue shales were deposited in the saline

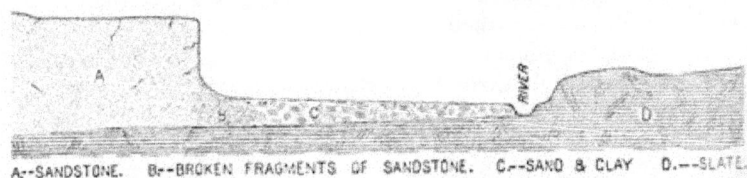

A.--SANDSTONE.   B.--BROKEN FRAGMENTS OF SANDSTONE.   C.--SAND & CLAY   D.--SLATE.

Fig. 2.—Geologic structure of the Yellowstone Valley at Billings.

waters of an inland basin where lime, sodium, and magnesium salts were deposited in large quantities. If there were any alkaline carbonates or chlorides present at that time chemical reactions have taken place and the carbonates and chlorides have been changed or entirely removed from the soil, as there is little or no trace of these present except in the form of carbonate of lime. The vast quantity of carbonate of lime present indicates the possibility of the previous existence of sodium carbonate in some quantity, but if it did exist there has been a reaction between the gypsum and the sodium carbonate whereby calcium carbonate and sodium sulphate have been formed, as there is but a trace of sodium carbonate in the rocks or in the soils of the valley.

A number of illustrations (Pls. II, III, IV, and V) are given showing in general and in detail the character of the sandstone and shale bluffs which border the valley and from which the soils of the valley have been derived. The legends on the illustrations will show quite plainly the object which they are intended to represent.

The weathering of these two rocks has given rise to the soils of the valley. From what has previously been said it will be seen that the rocks themselves are the present source of the soluble salts in the valley. As the rocks weather, a portion of the soluble salts is removed in the springs and seepage waters, but the removal is not nearly so complete as is the case in the humid portion of the United States, because the small rainfall renders the escape of all of the excessive amounts of salts impossible.

The two types of rock give rise to two distinct types of soil in the valley—one a sandy soil, derived from the disintegration of the sand stone rock, giving a soil of open texture easily worked, in which there is less trouble from alkali on account of the more perfect drainage and less risk of the accumulation of seepage waters; the other type is a stiff clay or gumbo formed from the disintegration of the shales. These shale soils are extremely fertile when in good condition, but are quite difficult to work. They are easily puddled and are rendered almost impervious to water by the excess of the soluble salts which they usually contain, and it is upon these soils, with their poor underdrain-age, that the greatest amount of trouble has arisen from the accumula-tion of seepage waters and salts in the overirrigation of the soils in the valley.

Between these two extremes of sandy soil and gumbo, in areas where the layer of sandstone has not been completely removed, the soils are blended in all possible combinations, from the pure type of the sandy soil to that of the gumbo.

The following table gives the mechanical analyses of a number of soils from Billings, which indicate the difference in the texture of the soils which has been noticed.

Mechanical analyses of soils.

No.	Locality. (Miles from Billings.)	Description.	Moisture in air dry sample.	Organic matter.	Gravel (2 to 1 mm.)	Coarse sand (1 to 0.5 mm.)	Medium sand (0.5 to 0.25 mm.)	Fine sand (0.25 to 0.1 mm.)	Very fine sand (0.1 to 0.05 mm.)	Silt (0.05 to 0.01 mm.)	Fine silt (0.01 to 0.005 mm.)	Clay (0.005 to 0.0001 mm.)
			P. ct.	P. ct.	P. ct.	P. ct.	P. ct.	P. ct.	P. ct.	P. ct.	P. ct.	P. ct.
3756	2½ N	Sandstone bluff soil.	1.22	2.66	0.00	0.00	0.17	29.39	52.34	3.29	0.88	9.65
3322	11 W	Silty type, creek soil.	2.98	4.40	0.00	0.00	0.16	7.96	28.79	34.45	4.67	17.25
3309	5½ W	Sandy gumbo	1.56	4.66	0.00	0.00	0.20	11.72	45.05	14.69	3.49	19.90
3308	5½ W	do	1.94	3.30	0.00	0.10	0.46	15.64	39.59	14.62	3.58	21.30
3307	5½ W	do	2.35	3.72	0.00	0.02	0.32	21.37	38.27	6.99	3.15	22.55
5306	3 W	Gumbo	3.20	3.50	0.01	0.40	1.58	20.40	27.67	11.71	4.02	27.30
3769	5 W	Heavy gumbo	3.74	4.22	0.04	0.03	0.19	11.65	24.63	15.15	4.40	35.55

The first sample is a very pure type of sandstone soil taken from the top of the bluff about 2½ miles north of Billings and was derived from the decomposition of the soft layers of fine sandstone which cap the bluffs. These soils are very light and loose and have very free underdrainage. As a matter of fact, they leach readily, and, although they afford the best possible conditions for irrigation in that seepage waters are not likely to accumulate in them, it is probable that they would not last very long, as the soluble salts would easily and quickly be removed from them. Soils of this type are found in many parts of the valley, and there is little or no danger from seepage waters or from an accumulation of soluble salts at the surface, although they contain considerable quantities of such salts at depths below the surface.

The other samples in the table are seen to grade up through the mixed sandy gumbo to the pure form of gumbo with from 27 to 35 per cent of clay. The mixed soils are the most abundant in the valley.

METHOD OF DETERMINING THE SOLUBLE SALT CONTENT OF SOILS.

The electrical method of determining the soluble salt content of soils has been described in Bulletins 8 and 12 of this division. Briefly, the method consists in taking a sample of soil with an auger at any desired depth in the field, adding sufficient distilled water to thoroughly saturate it and bring it into the condition of a thick paste. This is then filled into a hard rubber cell with metal electrodes, and the electrical resistance of the saturated soil determined with a modification of the apparatus described in Bulletin No. 8, a full description of which will be published at a subsequent time.

The resistance so found can be taken as the resistance of a salt solution filled with inert grains of soil. The effect of the soil grains is to increase the resistance about 100 per cent. This effect is constant for all ordinary soils. The resistance of the soil moisture can be calculated by multiplying the resistance of the cell by this factor and reducing for temperature, which has previously been determined. This will give the resistance of an amount of salt solution equal to that in the cell and without the presence of the soil. The specific resistance of this solution can then be determined, and by comparing this with the specific resistance of solutions of different strength of sodium chloride or of any other salt, the actual amount of salt can be determined in terms of the salt used as a basis for comparison. In this investigation the actual composition of the alkali, as determined by Dr. Traphagen, was taken as the basis of calculation, and all of the results have been worked out in terms of this analysis, which is given on page 10.

Plate VI shows the complete outfit for the determination of the soluble-salt content of soils as actually used in the field. The sample is taken with the auger, mixed in the field in a porcelain dish with distilled water, filled into a hard-rubber cell, and the resistance is then taken. The temperature of the soil in the cell is taken. The calculations can be made there or at any subsequent time. If the amount of moisture added is once actually determined, the variations in the amount added in different experiments have very little value. The moisture determinations do not, therefore, have to be made in each case.

This method is extremely sensitive and is very rapid. The determinations can be made by one man quite as fast as the samples can be drawn by another, and much faster if the ground is hard and dry.

Bul. 14, U. S. Dept. Agr., Div. Soils

PLATE VI.

FIELD APPARATUS USED IN SALT DETERMINATIONS.

Where the character of the soils vary considerably throughout the region or at different depths, it would require, of course, frequent chemical analyses to give a basis for the calculations; but over the area examined at Billings Dr. Traphagen had already shown that the alkali was very uniform in its composition, and no such frequent examinations were considered necessary.

PLAN FOR THE INVESTIGATION.

The objects of the investigations may be briefly stated. It was important to know the amount of soluble salts in the principal types of soil, which had been kept under approximately the same conditions as to exposure, cropping, and irrigation. It was important to determine the amount and distribution of the salts above the ditch where the land had never been irrigated and in the irrigated districts, and it was furthermore important to study the distribution of the salts in soils which had been or were now being ruined by the presence of alkali.

To carry out this plan three lines of borings were run—one of 5 miles in length and the others, for more detailed study, of about 1½ miles and one-fourth mile, respectively, in length. The longer section began above the ditch and went down toward the river; the others extended from an alkali flat and from a drainage ditch back into the higher levels. The position of these sections is shown on the sketch map of the valley. In each of these sections a number of borings were made and the salt content determined at every foot in depth down to 10 or, frequently, 15 feet. These borings were all numbered and their position accurately marked on the working maps. In addition to this, a section or square mile of land was studied in great detail and borings were made at frequent intervals to a depth of 10 or 15 feet. A number of special borings were also made to study the relation of the different types of soil to the amount and distribution of the salts.

The results of the investigations have been illustrated in the accompanying diagrams.

One set of the diagrams (Pls. VII, VIII, and IX) illustrates graphically the salt content of the soils found in the sections that were run; other charts (Pls. X, XI, XII, and XIII) show graphically the depth to standing water and the amount and distribution of salt for several depths in the section of land which was plotted. This gives an underground map showing the depth to standing water and the distribution of the alkali salts at several levels below the surface. Such a map is invaluable for the thorough understanding of the conditions under the soil with which the planters have to deal. If danger threatens, it shows the direction from which it is to come; where the land has already been injured, it shows exactly what the conditions are and where the cause is located. Furthermore, the method of salt determination is so sensitive and so rapid that inquiries of this kind can be readily extended over large areas with comparatively little expense.

As a result of the investigations at Billings it was found that plants could just exist with 0.45 of 1 per cent of the soluble salts present, equivalent to about 15,000 pounds per acre-foot, and this is taken as the limit of plant production. The soluble salt content of soils in the humid portion of the United States ranges from 50 pounds per acre-foot in the sandy soils of the Atlantic coast to as much as 3,000 or 4,000 pounds in some of the heavier agricultural soils. The average amount would be considerably less than 1,000 pounds per acre-foot.

THE RAINFALL AND SEEPAGE.

There are no available records of the amount of rainfall at Billings, but at Miles City the Weather Bureau records show an average annual rainfall of 12.8 inches. From May to September, inclusive, there are 6.7 inches, and during July and August 1.3 inches.

Plate XIV shows a section of the soil at Billings from a cañon just below the ditch. The whole depth of the face of the exposed soil is about 20 feet. About halfway down, or at a depth of 10 feet from the surface, there appears to be a layer of water-bearing sand and gravel through which a continuous slow seepage of water takes place. This probably came from the canal, which was perhaps one-fourth of a mile away. Above this point the soil appears to be quite dry. In September, 1897, when this photograph was taken and the water was actually running, a boring was made about a mile above the ditch. The upper soil appeared to be air-dry, but about 3 feet below the surface the soil was perceptibly moist, although there had been no rain for three or four months at that time. It was stated that standing water could be found at this place at a depth of 20 or 30 feet below the surface, which would correspond with the depth of the strata from which the water was running in the cañon below the ditch. We have then the problem of a desert with standing water at a depth of from 20 to 30 feet from the surface overlaid with 20 feet of almost air-dry compact earth and with only 12 inches of annual rainfall which appears to be insufficient to soak down to this strata.

When the rains occur in the spring and wet the surface of the ranges, vegetation flourishes in the most luxuriant way and the grasses give very good grazing. The rains, however, appear to be only sufficient to wet the surface to a very slight depth, and the water is quickly used up, and true desert conditions prevail during the summer time. From all the evidence it does not seem probable that the rainfall reaches down into the water-bearing strata to any great extent. It is not probable that the water-bearing strata is supplied from the local rainfall. It appears that in the dry season the soil is moist from 3 feet down, but so slightly moist and the depth of the dry material is so thick that it is altogether unlikely that the spring rains pass down to any appreciable extent locally in any one year. The conditions, therefore, are unfavorable to a natural leaching of these soluble salts except through the exceedingly slow movement there may be in the slightly moist subsoil.

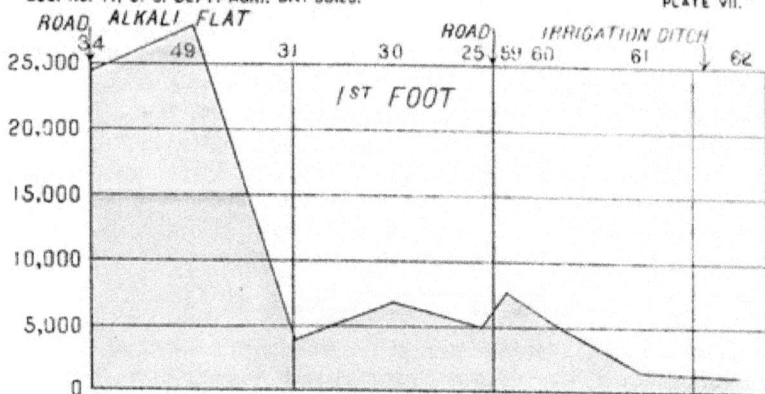

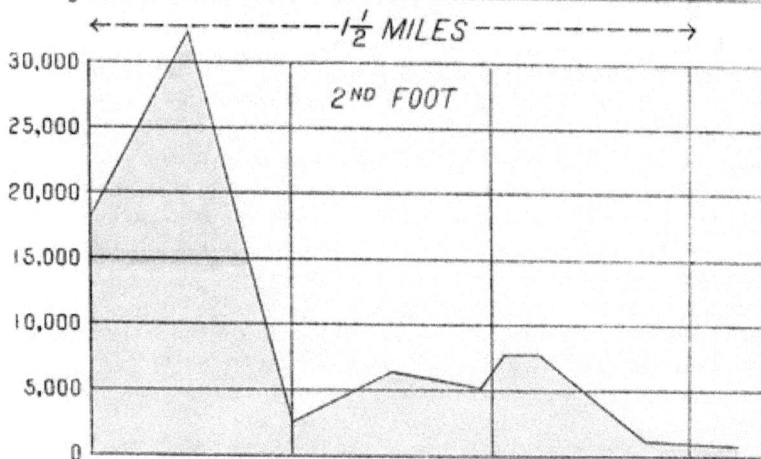

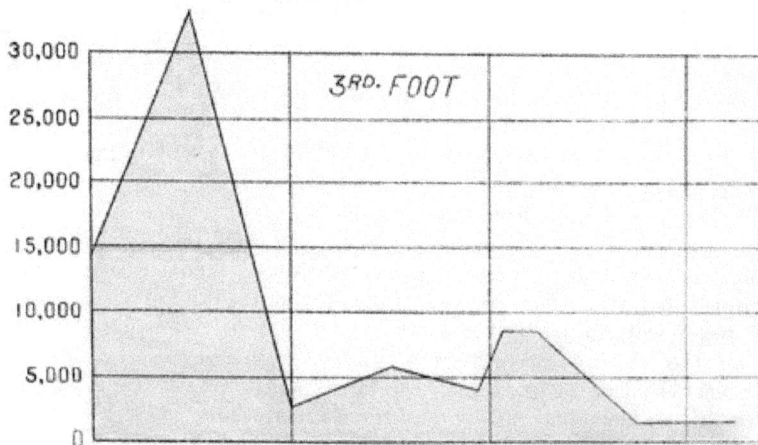

SECTION B. THROUGH BORINGS 34 TO 62.

SALT CONTENT FROM A LINE OF BORINGS ONE AND A HALF MILES OUT FROM AN ALKALI
FLAT. THE BLUE COLOR INDICATES THE EXCESS OF SOLUBLE SALTS.

SALT CONTENT OF THE SOIL.

An investigation was made of the salt content of a marked type of the sandy soil from above the ditch upon which water had never been applied through methods of irrigation.

The following table shows the amount and distribution of the salts in two places:

Amount of soluble salt in sandy soil.

UNIRRIGATED.

Depth (feet).	Boring 62.		Boring 64.	
	Per cent of salt.	Pounds per acre-foot.	Per cent of salt.	Pounds per acre-foot.
0- 1	0.033	1,155	0.042	1,470
1- 2	.019	665	.011	1,435
2- 3	.045	1.575	.005	1,925
3- 4	.027	945	.038	1,750
4- 5	.032	1.120	.045	1,575
5- 6	.028	980	.055	1,925
6- 7	.019	665	.056	1,960
7- 8	....	....	.170	5,950
8- 9	....	....	.238	8,330
9-10	....	....	.243	8,505
10-11	....	....	.205	7,175
11-12	....	....	.120	4,200
12-13	....	....	.163	5,705
13-14	....	....	.228	7,980
14-15	....	....	.178	6,230

It will be seen that there is not sufficient soluble matter down to a depth of 15 feet to prevent the growth of agricultural plants, as the amount does not approach the limit of 0.45 per cent, or 15,000 pounds per acre-foot. It is interesting to note, however, that the amount of soluble salt in the upper 7 feet of the soil is particularly small. There is 50 per cent more, perhaps, than is ordinarily found in the soils of the humid region. Below a depth of 7 feet, however, the amount of salt is considerably increased. It would appear that there were evidences here of a slow downward movement of soil moisture, and that under these constant conditions of slow seepage the amount of salt in the upper layers of the soil was constantly diminishing. No examinations could readily be made of the soil below a depth of 15 feet, as this was the extreme length of the auger, but from information furnished by some well-diggers thoroughly familiar with the locality it was learned that the soluble salt content increases below this to very large proportions. White layers strongly impregnated with salts are said to be found below this depth. The water from the wells contains too much of the salts to be of use for domestic purposes, although it is not so strong as to be harmful to cattle.

It is quite reasonable to suppose that the soluble salts had originally been uniformly distributed throughout the upper layers of these soils, and that from storm waters and the slow seepage of the slight amount of moisture which the subsoil contains the soluble material has been washed down from the upper layers. Certain it is that similar results follow from the first effect of irrigation where there is good under-drainage, as generally prevails in this sandy soil.

Borings were made in the same character of soil below the ditch and only a short distance away. The following table gives the salt determinations in three borings made in this irrigated sandy soil:

Salt determinations in irrigated sandy soil.

Depth (feet).	Boring 26.		Boring 27.		Boring 28.	
	Per cent of salt.	Pounds per acre-foot.	Per cent of salt.	Pounds per acre-foot.	Per cent of salt.	Pounds per acre-foot.
0- 1	0.046	1,610	0.038	1,330	0.033	1,155
1- 2	.049	1,715	.045	1,575	.037	1,295
2- 3	.052	1,820	.044	1,540	.028	980
3- 4	.066	2,310	.051	1,785	.030	1,050
4- 5	.097	3,395	.060	2,100	.048	1,680
5- 6	.106	3,710	.051	1,785	.048	1,680
6- 7	.128	4,480	.064	2,240	.048	1,680
7- 8	.147	5,145	.070	2,450	.047	1,645
8- 9	.112	3,920	.049	1,715	.047	1,645
9-10	.112	3,920	.049	1,715	.047	1,645
10-11	.056	1,960	.072	2,520	.044	1,540
11-12	.056	1,960	.072	2,520	.044	1,540
12-13	.058	2,030	.072	2,520	.044	1,540
13-14	.058	2,030	.072	2,520	.044	1,540
14-15	.058	2,030	.072	2,520		

It will be seen in this case that the amount of soluble matter, even to a depth of 15 feet, is comparatively small, and the amount throughout the whole depth is quite uniform. This indicates very strongly that the salts have been leached out of the soil and carried off in the underground drainage waters. The examination of the water in a well situated in this irrigated area of the sandy land gives additional proof that some of the salts have been removed. There is a well at the southwest corner of section 2, near boring 44, in which the water contains 0.119 per cent of soluble matter, or 60 grains per gallon. Other wells throughout the irrigated area frequently contain as much as 0.4 per cent of salts.

Where this sandy soil is overirrigated, or where from some physical cause the subsoil is compact and the drainage is poor and water accumulates within the subsoil, the seepage waters which move rapidly accumulate and frequently come up to the level of the surface of the ground. Under these conditions excessive evaporation sets in from the surface and salts accumulate until the soil moisture is so saturated that the salts are deposited at the surface as a crust. Such alkali flats are frequently found on the low levels of the Billings area, even in these sandy soils.

The following results indicate conditions in one of these alkali flats, where the sand is underlaid at a slight depth with a heavy gumbo subsoil:

Salt determinations in an alkali flat in a sandy soil.

Depth (feet).	Boring 49.		Boring 52.	
	Per cent of salt.	Pounds per acre foot.	Per cent of salt.	Pounds per acre foot.
0-1	0.792	27,720	0.229	8,015
1-2	.920	32,200	.191	6,685
2-3	.944	33,040	.182	6,370
3-4	.792	27,720	.175	6,125
4-5	.519	18,165	.159	5,565
5-6	.519	18,165	.213	7,455
6-7	.357	12,495	....	....
7-8	.357	12,495	....	....
8-9	.292	10,220	....	....

It will be seen that in boring 49, which was in the midst of the alkali flat, there was an excessive accumulation of alkali, beyond the limit of any agricultural plant, at least to a depth of 7 or 8 feet. Below this it rather looks as though the amount of salts was diminishing and that if the boring could have been carried deeper the salt content would perhaps have grown less.

In both these borings standing water was found between 1 and 2 feet of the surface of the ground. Evaporation had been going on for a number of years, the seepage waters being supplied by the overirrigation of lands on higher levels. At the present time the soil in boring 52 does not contain an excessive amount of alkali for the alfalfa, but the level of standing water is so near the surface that the roots of the plants are submerged and the crop can not be successfully grown. This is the first stage in the ruin and devastation that is being wrought, and boring 49 shows the final and complete stage when the land is given up to water and alkali.

When land is in the condition of boring 52, and before any notable accumulation of crust has appeared upon the surface, the land becomes covered with a heavy growth of weeds. All agricultural crops have ceased to grow for some time, and the land has been left out as a barren waste. Such a condition as is shown by the growth of weeds is usually thought to mean that the alkali is disappearing or is being used up by the weeds themselves and that the soil is again becoming fit for crops. In some cases this may be true, if there is sufficient drainage to carry off the excess of seepage waters; but in many cases the conditions simply indicate that the weeds are a class of plants which can thrive on wet ground and grow for a while luxuriantly. If the methods of irrigation are kept up and the seepage waters continue to collect and evaporate for a few years longer, alkali accumulates in sufficient quantities to kill even the growth of weeds, and the land truly presents the appearance of a desert.

Some of the planters are much inclined to accept the prevailing conditions and to look for crops like the Australian salt grass, which will grow in these alkali flats and wet soils. It is wrong to accept these conditions, however, and depend upon these makeshifts, when the conditions ought in the first place to have been prevented and ought now to be removed by radical and energetic methods of drainage or through better and more careful methods of irrigation.

From all the facts thus far observed it can be said that the first harmful effect observed in these sandy soils is caused by an excess of water, and if this is not immediately lessened further damage will result from the accumulation of soluble salts. If the excess of water is soon removed no permanent damage will result.

The heavier type of soil—that is, the gumbo soil—was shown to be derived from the disintegration of the shale. The still undecomposed shale in the bluffs on the south side of the valley was found to be penetrated in every direction with veins of gypsum and the soft shale itself to be permeated with large quantities of sodium and magnesium sulphates. The soils resulting from the disintegration of the shale form a heavy, sticky, blue clay quite impervious to water. The drainage is so slow through this fine, impervious material that large quantities of the salts remain in the soil.

On account of the poor drainage and the slow movement of the subsoil waters through this material there is great danger of overirrigation, and the problem of irrigation, which is easy on the well-drained, sandy lands, becomes far more complicated and much more difficult to manage on these heavy gumbo soils. Great care has to be taken, not only in the application of water, but in the actual cultivation of these soils, as they are liable to be ruined for a time at least for alfalfa. On account of the large water content, the fineness of the particles, and the amount of salts these soils contain they easily puddle, and if they are worked when too wet clods form and it is very difficult to reduce the field again to a good tilth.

On account of these properties the heavier or gumbo soils have to be farmed with very great care. Not only so, but there is great danger from seepage waters from neighboring plantations on higher levels. The soils themselves are naturally extremely fertile and very strong and last very well if well cared for. The following table gives the results of salt determinations in typical gumbo soil above the ditch, which has never been irrigated.

The salt content of a heavy unirrigated gumbo soil.

Depth (feet).	Per cent of salt.	Pounds per acre-foot.
0-1	0.065	1.225
1-2	.068	1.330
2-3	.054	1.690
3-4	.200	7.000
4-5	.333	11.655
5-6	.347	11.795
6-7	.253	8.855
7-8	.253	8.855
8-9	.262	9.870

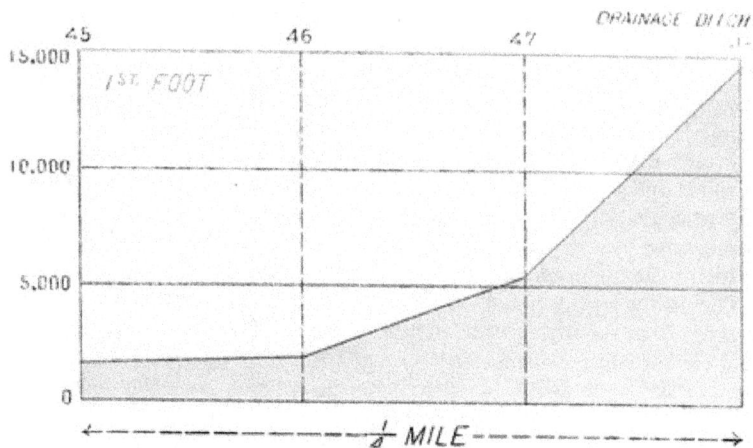

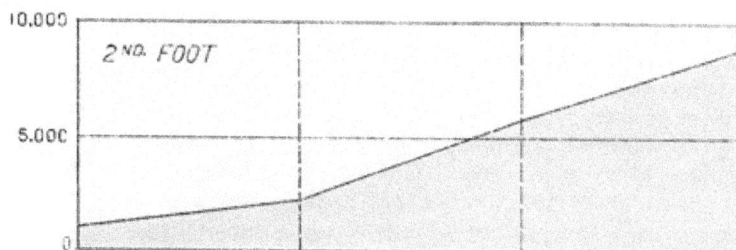

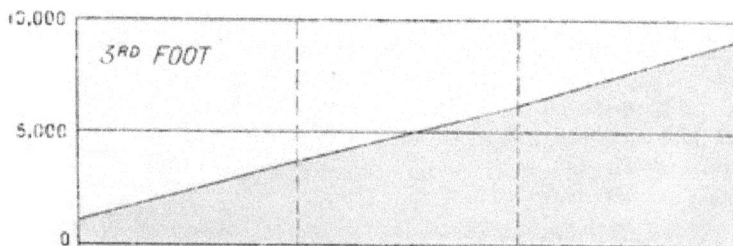

SECTION C THROUGH BORINGS 45 TO 48.

SALT CONTENT FROM A LINE OF BORINGS EXTENDING ONE-QUARTER MILE BACK FROM
A SMALL DRAIN IN AN ALKALI FLAT.

It is apparent from the table that there is a considerable quantity of salt at a depth of 5 feet and from there down. This boring, with others, is illustrated graphically in fig. 3.

When water is applied to this gumbo soil in the practice of irrigation, the first effect is to reduce the amount of soluble salts in the upper

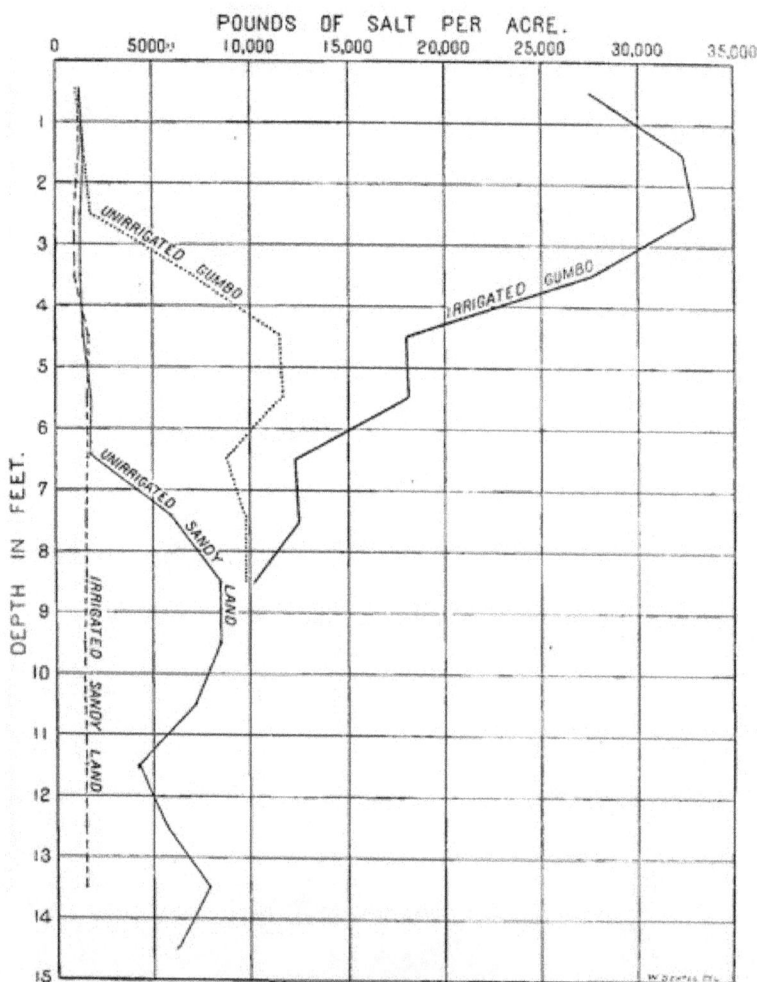

Fig. 3.—The salt content of sandy land and gumbo, with and without irrigation.

layers of the soil. If there is good drainage this excess of salt may be removed altogether from the soil. If the drainage is slow and inefficient, however, or in case of a large excess of seepage waters from higher levels, the water soon accumulates in the subsoil and quickly rises and drowns out all vegetation. If the conditions remain in this way for some time and evaporation is allowed to continue, enormous quantities of salt accumulate in the upper layers of the soil, and the land is finally turned out as an alkali flat.

The accompanying table shows the salt content of an alkali flat in this gumbo soil to a depth of 6 feet.

Amount of soluble salts at different depths in overirrigated gumbo land.

Depth.	Boring 36.	
	Per cent of salt.	Pounds per acre-foot.
0–1 inch..	1.023	a 2,975
0–1 foot..	.757	24,710
1–2 feet..	.714	34,570
2–3 feet..	.634	21,105
3–4 feet..	.612	21,035
4–5 feet..	.589	20,905
5–6 feet..	.187	6,650

a Pounds in acre-inch equivalent to 35,700 pounds in acre-foot.

It will be seen that the salt has accumulated to enormous proportions in the top 5 feet of the soil. The conditions show that the solution is so strong that a white crust is formed over the surface. However, on account of the absorptive powers of the soil, the solution immediately under this crust and in contact with the soil was only 3 per cent, notwithstanding the fact that the salt ordinarily is very soluble in water. A solution of this strength, however, is entirely too strong for any cultivated crop, and the alkali flat presented a very desolate appearance, as seen in Plate XV.

A line of borings was run from the center of this alkali flat for a mile and a half back to the main canal. A number of borings were made to a considerable depth and salt determinations calculated for every foot in depth in each of these borings. The result of these investigations for the top 3 feet is shown graphically in Plate VIII. In this illustration the span bounded by the curve above the line of 15,000 pounds per acre represents the relative area in which there is an excess of alkali and the amount of this excess. It will be seen that adjoining the canal and for two-thirds of the way down to the alkali flat there is but little alkali, as though the irrigation water had removed the salts from this portion of the land and that they had then accumulated in the alkali flat, which is at a somewhat lower level. These irrigation waters slowly seep through the underground channels down into the natural drainage system, which is represented in this case by the alkali flat on account of its somewhat depressed condition. The salts first appear in these low places in the line of underdrainage and as the evaporation of the water goes on the salts accumulate, gradually extending up and enlarging the alkali flats as the water rises until the level of the surrounding area is reached, when the whole district is abandoned. Along the line represented in this diagram the area around and back from the alkali flat is first-class alfalfa land and the property is considered very valuable, but only three or four years ago the alkali flat itself was considered just as valuable, and the alarming feature of the whole thing is that the owners know that, if these conditions continue, the alfalfa field itself will be ruined and will have to be abandoned in a few years unless very energetic means are taken to arrest the progress of the trouble. The chances are, of course, that this condition has not arisen from the local application of water at this place. It is possibly a result of injudicious methods of irrigation on the adjoining lands at higher levels. One of the most discouraging features of the whole problem is that the owner of such a tract of land may use the most approved methods of irrigation, and yet be completely ruined by the excessive and injudicious use of water by his neighbor, who may himself escape the injurious effects of his own crude methods, at least for many years after his neighbor has been ruined. In the contemplation of such a problem as is presented in the Yellowstone Valley, therefore, there are certain property rights that may easily be abused, causing very disastrous results to appear upon a neighbor's property. It makes the whole problem very difficult to deal with, especially as it would be extremely difficult to show the source of the trouble and to locate the offending person. It is a problem, however, which will have to be taken up; and if the property owners do not themselves take adequate care of their drainage systems and use intelligent methods of irrigation, some means must be found of compelling them to do so, or to give redress to their unfortunate neighbors.

This accumulation of salts is very harmful in the puddling effect on the soil. The flocculation of the soil grains is broken up and the grains are separated out into their most uniform position, where they offer the greatest possible resistance to the flow of water. This puddling can only be relieved by draining off the water and salts, and this drainage is rendered exceedingly difficult by their presence, so that the reclamation of these alkali flats on the gumbo soil is an exceedingly difficult, slow, and expensive undertaking.

The formation of these alkali flats is in a way an evidence of the effort of nature to correct the faults of our crude system of irrigation. The salts are being carried off into the natural drainage of the country, but the process is very slow and the excess of seepage water and the salts themselves collect in these places on account of the inability of the soil to let them pass as rapidly as the excess of water is supplied. This suggests the only feasible method of reclaiming these lands and, indeed, of preventing the accumulation of the salts which will occur except under the most careful and judicious methods of applying water. In cases like that under consideration, where the damage has already been done, the natural drainage is so slow that it does not afford adequate relief, and in fact is but a sign of impending ruin for a very much larger area. The only way to reclaim the land is to put in an efficient system of drains, preferably of underground tile drains. It is urged against this idea that the land is not worth the cost of the investment in putting in a system of tile drains. This, of course, is an economic problem which is entirely dependent upon conditions of market, transportation facilities, and other commercial considerations. It may or may not be profitable at this time to protect the lands from destruction and to reclaim those that have been destroyed. It may be cheaper to move off into new areas, but the time will come, if it has not already come, when the land in the Yellowstone Valley and in similar situations will be worth the care and expense necessary to protect it from ultimate destruction. The amount of money now invested in the Yellowstone Valley is enormous, and the continuance of prosperity is entirely dependent upon the care which is taken in the methods of irrigating the lands. Property worth thousands of dollars may be ruined in a few years and become utterly worthless. The experience in the valley shows that this has been the case in the past, and there is much uneasiness felt in regard to large areas which show signs of the rapid spread of alkali. (See Pl. XV.)

There is abundant evidence that thorough underdrainage will reclaim these lands, and if introduced in time will prevent any such disastrous results as those which have been described. There has been no thorough system of tile drainage tried, but a few efforts have been made to reclaim the abandoned lands by open drains. That these have been efficacious the following investigations will show:

A drainage ditch had been dug in the alkali flat on section 2 and the excess of water had been continuously removed for some time before this investigation was made. A line of borings was made from this ditch back about one-fourth mile to see how the salt content had been changed. The results of the determinations for the top 3 feet is shown graphically on Plate VIII, and the actual determinations are given in the accompanying table, which represents the per cent of soluble salt found at different distances from this ditch.

Salt determinations at different distances from a drainage ditch.

Depth (feet).	Boring 45.	Boring 46.	Boring 47.	Boring 48.
	Per cent.	Per cent.	Per cent.	Per cent.
0-1	0.047	0.054	0.155	0.419
1-2	.030	.066	.164	.253
2-3	.036	.142	.177	.257
3-4	.031	.103	.213	.238
4-5	.045	.169	.191	.275
5-6	.045	.120	.191	.307
6-7	.043	.136	.295	.273
7-8	.043	.162	.237	.337
Average	.040	.108	.194	.290
Total pounds per acre 8 feet deep	11.200	30.280	54.320	81.200

Boring 46 shows the amount of salt about 300 yards from the drainage ditch. The next column shows the amount at about 100 yards from the ditch, while the last column shows the amount closely adjoining the ditch. It will be seen from this table and from graphic representations

PLATE IX.

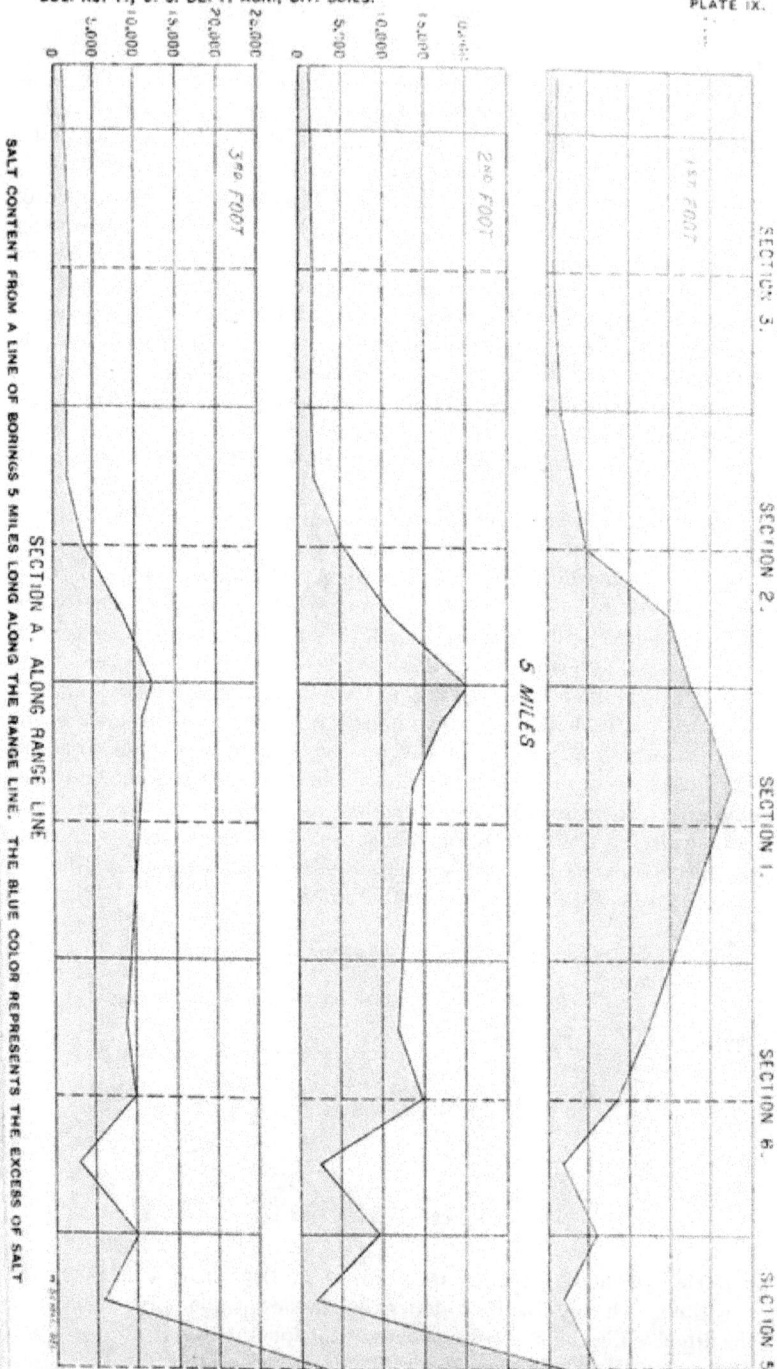

SALT CONTENT FROM A LINE OF BORINGS 5 MILES LONG ALONG THE RANGE LINE. THE BLUE COLOR REPRESENTS THE EXCESS OF SALT

SECTION A. ALONG RANGE LINE

(Plate VIII) how the salts have accumulated at the lowest point of the drainage system and how they are being removed by the drain as the water from the irrigating ditch seeps down and the excess of seepage water is carried off. The amount of salt found at the drainage ditch is already below the 15,000 pounds limit of crop production, while farther out in the alkali flat, under presumably the same conditions as existed at the drainage ditch when it was cut, the amount of salt is upward of 35,000 pounds per acre-foot.

It is easy to show in another way the beneficial effects of under-drainage. The land around Billings is underlaid at a depth of from 6 to 8 feet by a layer of gravel. Within the last few years it has been necessary to construct a ditch around the town of Billings to cut off from the town the seepage waters from the irrigated lands. This ditch is cut to a depth of 6 or 8 feet, so that it is in the gravel through its whole length, and it receives the water from two or three natural draws and all of the seepage water from four or five sections of land. No sewage is allowed to flow into the ditch from the town.

A number of observations were made upon this ditch during the month of June. It was estimated that the average flow of water during this time was about 40 cubic feet per second. Frequent determinations were also made of the soluble salt content of the water flowing in this drainage ditch.

The following table gives the amount of salt found during the month of June:

Examination of drainage water.

	Salt per cent.	Tons of salt removed per hour.
June 4	0.302	13.6
June 6	.424	19.1
June 7	.422	19.9
June 11	.449	20.2
June 12	.263	12.7
June 13	.481	21.6
June 16	.227	10.3
Average	.370	16.6

It will be seen that the ditch is doing a great work in removing the salt content from the overirrigated lands around the town. At the rate at which the salt was being removed at the time these observations were made, the ditch was removing about $16\frac{1}{2}$ tons per hour. If this rate was continuous, it would drain 1 per cent of salt from the upper 5 feet of about 900 acres of land per year. As a matter of fact, while the ditch rarely if ever stops flowing, the flow is not always as great as during the time of this investigation. Still the figures give some idea of the enormous results which may be accomplished by a judicious system of drainage in the reclamation of these alkali lands and the protection from an undue accumulation of salt and seepage waters.

Plate 1 (frontispiece) gives a distant view of the town of Billings from the slate ridge over toward the sandstone ridge on the other side of the valley. This portion of the valley has been almost swamped with seepage waters and ruined with the rise of salts. The white crust of alkali shows in the illustration as covering most of this portion of the valley for 3 or 4 miles from the town of Billings. It is through this area that the drain just mentioned has been started. It may be well to state that the water in the main irrigating canal (Pl. XVI) that is taken out of the Yellowstone River about 40 miles above Billings is very free from alkali, the water, therefore, being used for irrigating purposes, although carried for this great distance through the valley, was fresh and free from salts. From all sources, however, where this water was escaping as seepage water it is seen to be loaded with excess of salts, and where free to flow off readily the salts will be carried off instead of accumulating in the alkali flats.

Besides the examination of the drainage ditch around the town of Billings many determinations were made of the salt content of springs and wells. The following table gives the determinations of the salt content from a number of places.

Salt content of springs and wells near Billings.

	Per cent.
Spring on the sandstone bluff	.029
Well on west side of section 2	.119
Spring on west side of section 2	.212
Well on north side of section 2	.309
Spring on north side of section 6	.433
Spring in center of section 5	.437
Well on south side of section 1	.536
Well on north side of section 2	.538

Well water containing more than one-tenth of 1 per cent is ordinarily considered unfit for domestic use. In most of these wells the water was less than 5 feet below the surface of the ground, and the water tasted strongly saline.

It will be seen, therefore, from all of these sources that the water in the soil is charged with this excess of salt. If there is a ready means provided for it to leave the soil, there will be no excess of soluble salts. It would, of course, be unfortunate to depend upon this to carry off the salts from reckless overirrigation, for in removing these salts much valuable plant food may also be lost and the soil in a measure impoverished. It is necessary, therefore, even with a system of efficient underdrainage, to use great care, so that there shall not be more loss through underdrainage than is necessary. It is perfectly evident, however, on the other hand that if these conditions continue and the water rises closer to the surface than it is at present, that the seepage waters and the accumulation of salts together are likely to prove very disastrous over larger areas.

UNDERGROUND MAPPING OF THESE SOIL AREAS.

Having found the source of the alkali salts in the sandstone and slate rocks which border the valley from which the soils of the valley have been derived, and having determined the character of the soils resulting from each class of rocks, the texture and relation of these to the alkali salts, to underdrainage, and to seepage, the next step was to make a detailed examination of such an area as time would permit for the construction of an underground map, which should show the amount and distribution of the soluble salts at different levels below the surface. For this purpose a preliminary line of borings was made for a distance of about 5 miles along the range line, as indicated on the sketch map of the valley. Borings were made at frequent intervals and the amount of salt determined for every foot in depth. Plate IX shows graphically the amount of salt found in the top 3 feet of the soil along the range line. The line goes just beyond section 5 in the sandy prairie above the ditch, and continues throughout section 3 in the same character of soil, in which the amount of alkali is quite small and no larger than is ordinarily found in the soils of the humid regions. Beyond this the heavy gumbo soils prevail, and the seepage waters have accumulated and the salts have accumulated until the whole of section 1 and a part of section 2 are lying out as an alkali flat. This accumulation of salt and the accompanying accumulation of seepage waters which cover section 1 along this range line, is gradually extending and covering larger and larger areas.

A detailed examination was finally made of section 2, over which the ruin caused by the rise of seepage waters and the accumulation of salts was well advanced. The level of standing water was determined as well as the amount of salt to a depth of 10 or 15 feet. From this examination maps have been constructed, several of which are given in this bulletin, illustrating graphically the amount and distribution of salt in the top 3 feet of soil. The depth to standing water is illustrated on Plate X, with contour lines showing the depth in feet from the surface of the ground to the level of standing water at the time of the observations. The other plates represent with contour lines and with different tints different amounts of alkali salts. The green color on Plate X shows the area over which the depth to standing water is less than 2 feet from the surface and over which, for this reason, alfalfa will no longer grow. On the other plates, XI, XII, and XIII, the blue color represents the area containing over .15 per cent, or 15,000 pounds per acre-foot of soluble salts, which is taken as the limit of profitable crop culture. These plates are fully described in another place.

7769—No. 14——3

EXPLANATION OF PLATES.

PLATE X. This represents the depth to standing water on section 2, upon which alfalfa was gradually dying out. The green color represents the area where the depth to standing water is within 2 feet of the surface of the ground. Over this green area the alfalfa has died out and can no longer be grown. The figures and contour lines indicate the depth to water. The pink area is still in good condition for alfalfa. Much of the green area, especially in the lower section of the area, is a swamp, and efforts have been made to drain it in two places with open ditches. This accumulation of water is entirely due to overirrigation of the surrounding lands. Formerly this whole section of land was a valuable tract of fine alfalfa soil. Before irrigation was introduced into the valley standing water was probably not less than 25 or 30 feet from the surface.

PLATE XI. This illustration shows the amount of soluble salts in the surface foot on section 2. The amount of salt is indicated by contour lines, the figures showing the amount of salt in pounds per acre. The pink color is used where there is at present no excess of salt, and the blue color indicates the area in which there is a large excess. The cross-hatching further brings out these areas. It is interesting to observe that the excessive salt content of the soil has not yet spread out so as to cover the whole of the swamp area of the section. It will be seen from Plate XIII that very nearly three-quarters of section 2 is rendered unfit for alfalfa or cultivated crops of any kind, as there is standing water within 2 feet of the surface. Hardly more than a third of this swamp area as yet contains an excess of alkali, but the records for the past few years show that the alkali is increasing and spreading rapidly. It is interesting to note, furthermore, the apparent path of the salt as it occurs in what appears to be a trough, probably due to some peculiarities in the structure of the soil.

PLATE XII. This represents the salt content in pounds per acre in the second foot of section 2. This is colored with yellow to indicate no present excess of salt, and with blue to indicate the areas containing an excessive amount of salt. The figures, contour lines, and hatching show clearly the actual amount of salt over different portions of the area.

PLATE XIII illustrates the amount and distribution of soluble salt in the third foot in depth in section 2. The colors and hatching represent the same features as in the other plates.

31

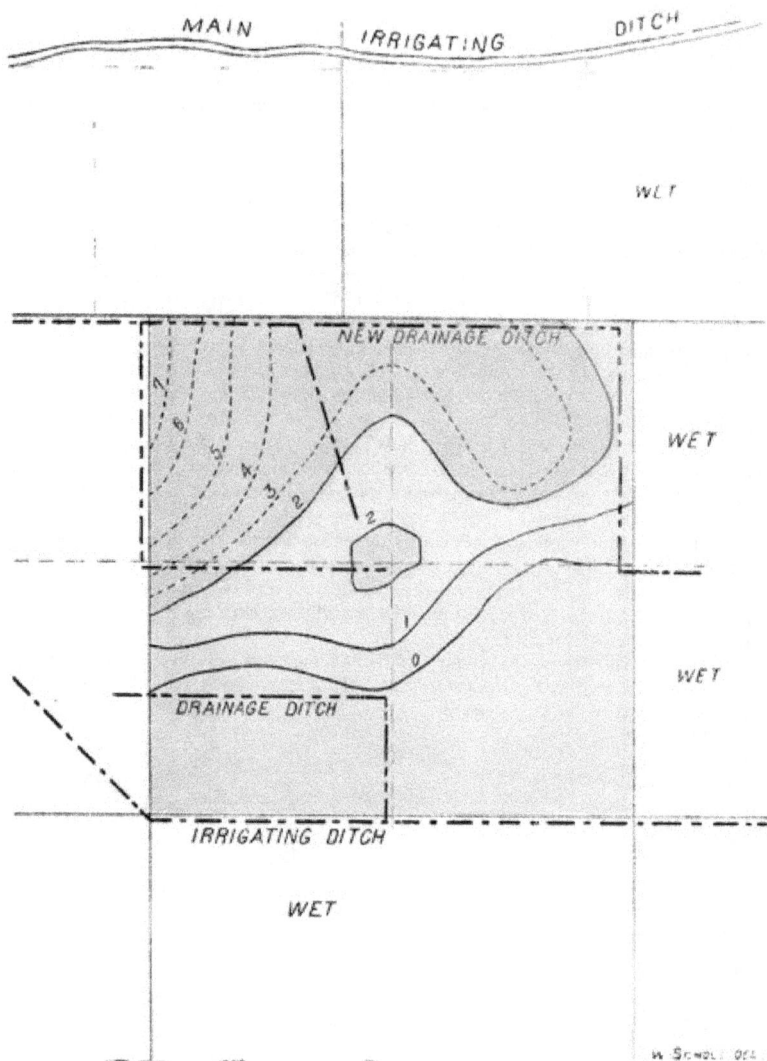

DEPTH IN FEET FROM SURFACE OF GROUND TO LEVEL OF STANDING WATER ON SEC. 2, T. 1 S., R. 25 E. THE GREEN AREA HAS STANDING WATER WITHIN 2 FEET OF THE SURFACE, AND ALFALFA WILL NO LONGER GROW HERE.

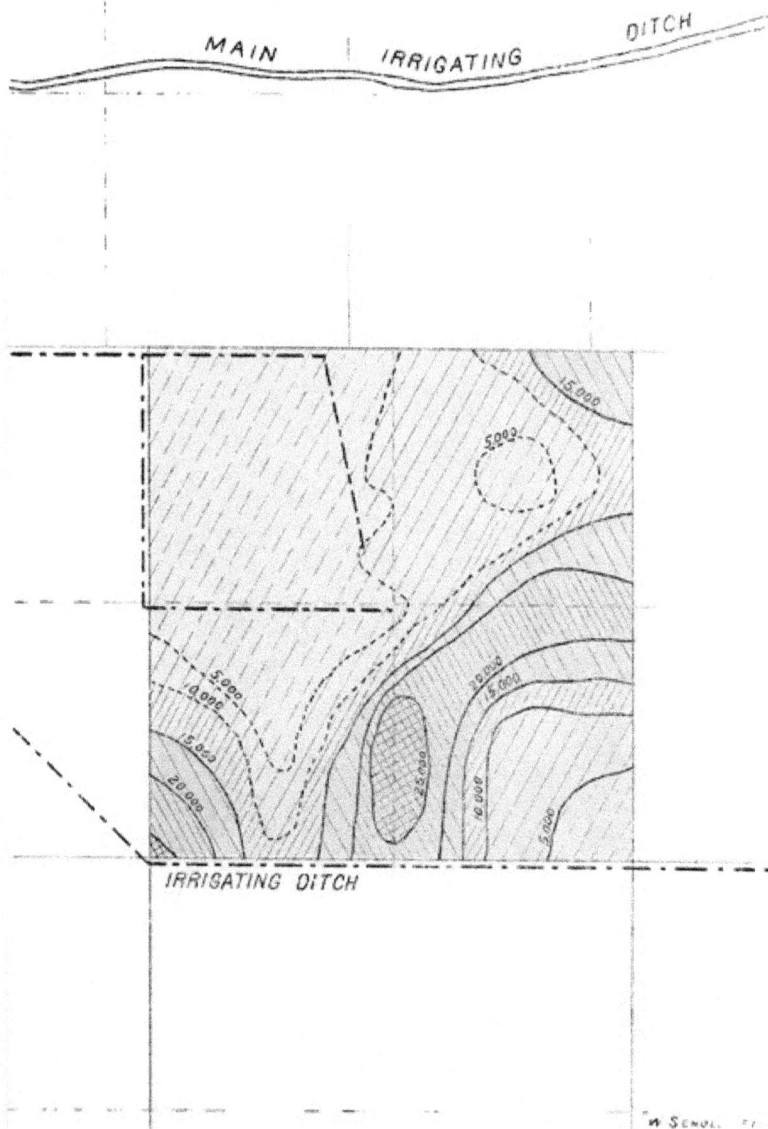

SOLUBLE SALT CONTENT OF THE TOP FOOT, SEC. 2, T. 1 S., R. 25 E. THE BLUE COLOR
INDICATES AN EXCESS OF SALTS.

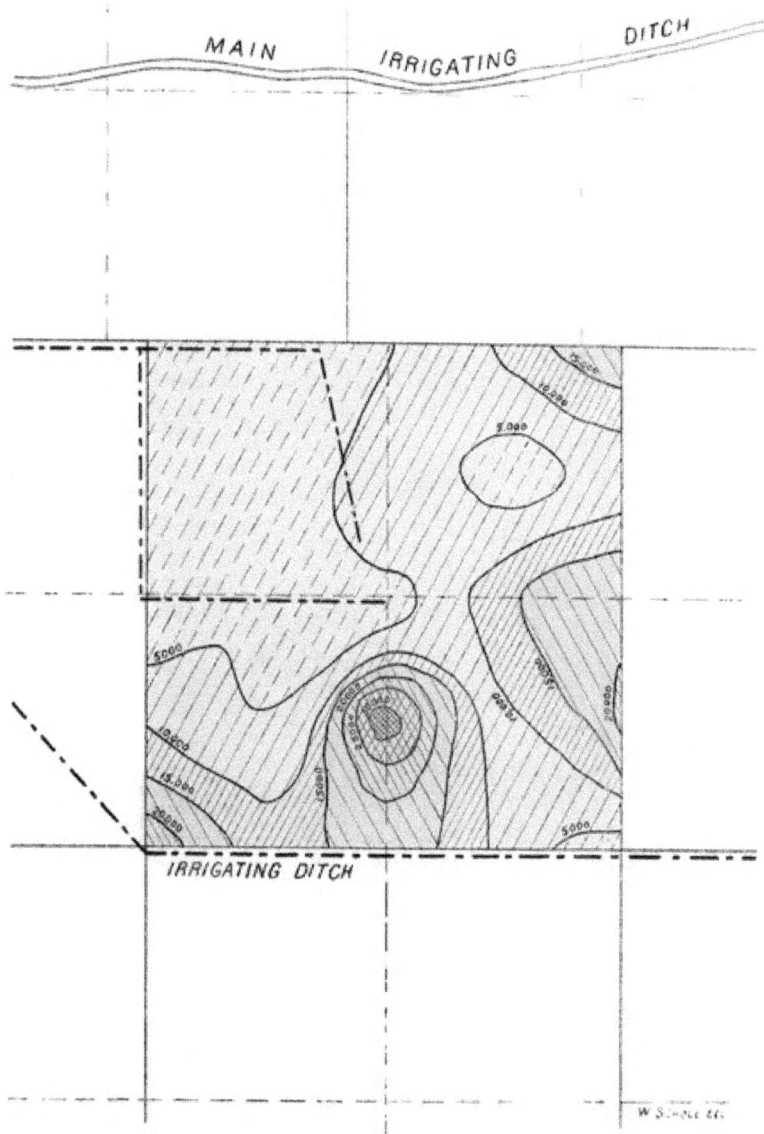

SOLUBLE SALT CONTENT OF THE SECOND FOOT, SEC. 2, T. 1 S., R. 25 E. THE BLUE
COLOR INDICATES AN EXCESS OF SALTS.

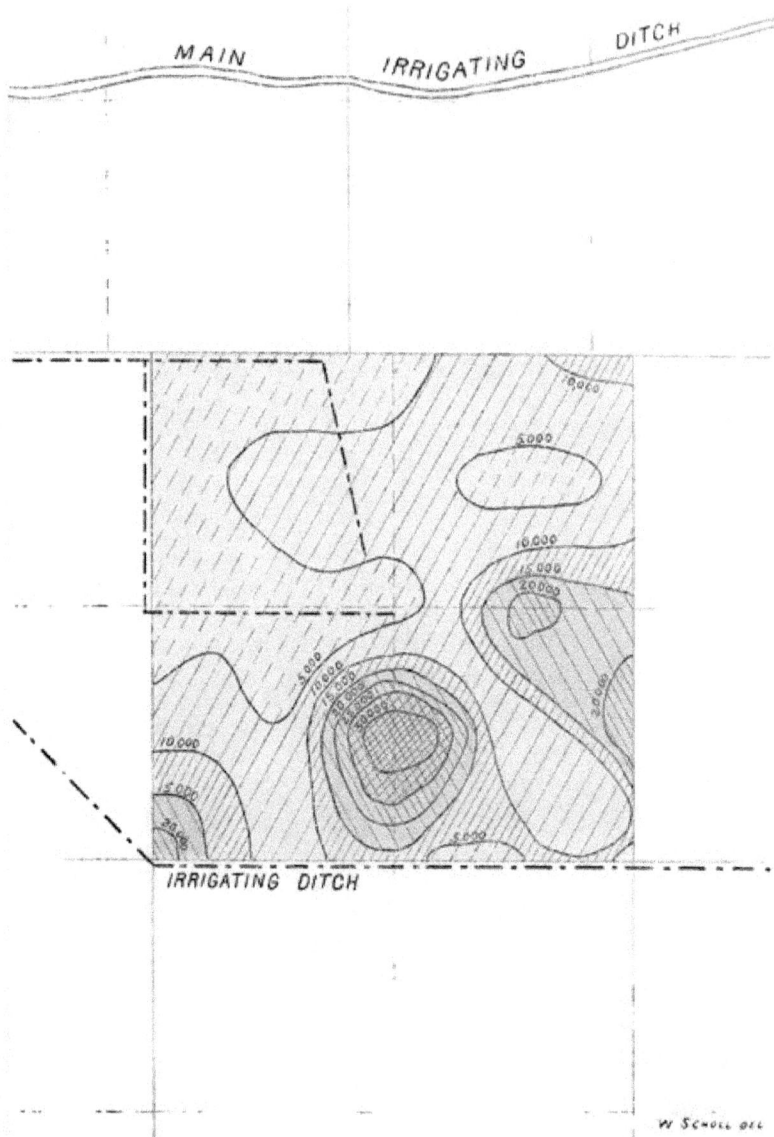

SOLUBLE SALT CONTENT OF THE THIRD FOOT, SEC. 2, T 1 S., R. 25 E. THE BLUE
COLOR INDICATES AN EXCESS OF SALTS.

PLATE XV

The value of such underground maps can hardly be overestimated to the owners of the land. It is seen just where the seepage waters and alkali salts are accumulating, from which direction they are coming, and just how drainage systems should be introduced to remove the trouble. They will show that some areas are quite safe for a number of years at any rate, and the maps will indicate other areas which will need careful attention or even energetic efforts in the prevention of or reclamation from damage.

SUMMARY OF THE INVESTIGATIONS AND CONCLUSIONS.

The results of these investigations show that the ultimate source of the alkali is in the sandstone, and particularly in the shale or slate rocks from which the soils have been derived. Before irrigation was introduced the salts were present in rather large amounts, but well distributed throughout the soil, and not in such large quantities as to be injurious to crops. The injury is due entirely to overirrigation, to the translocation and local accumulation of salts by means of seepage waters, and to the imperfect drainage facilities in the compact gumbo soils and the inability of the soils to remove the excess of salts and of seepage waters. The first trouble appears to be due to the seepage waters. This, of course, need not necessarily be so, but it appears to be the case in this locality. The open sandy lands, having better under-drainage, are not likely to be injured by a rise of salts except from an excessive application of water or in the low places in the path of the drainage system, especially when these are underlaid, as they are liable to be, by the heavy gumbo subsoils. The gumbo soil requires great care in cultivation, as it is easily ruined by the accumulation of seepage waters and the subsequent accumulation of salts. There are many areas in the valley, of course, which have still a low or moderate salt content which are probably safe for years to come. There are other areas in which the salts are now accumulating to such an extent as to render the future value of the land very uncertain, while there are still other areas which have gone beyond this stage, and what were once fertile tracts have been thrown out as barren flats. The investigations show, further, the very disturbing fact that the injury need not be due to a local application of water, but to the injudicious application of large quantities of it in remote localities and on neighboring farms over which the unfortunate person has no control and for the effects of which he has at present no redress.

The investigations point clearly to the natural methods of preventing this injury and of reclaiming the lands when once the injury has occurred. There is no question that the injury is due to the translocation and local accumulation of the salts which were formerly well distributed in the soils of the valley. Alkali has only been troublesome here after eight or ten years of irrigation. The trouble is always preceded by an accumulation of seepage waters, followed in a few years by the alkali incrustations on the surface of the land. This evidently points to the necessity of great care in the application of water in the methods of irrigation. This care must be exercised not only for the land which is being irrigated, but for the adjoining lands on lower levels. While a man can overirrigate a sandy tract with practical impunity to himself, he is likely to swamp his neighbor on a lower level. There are involved property rights, therefore, which will come to be recognized and which will have to be taken into consideration in any intelligent and safe system of irrigation.

Where the damage has been done, or where the conditions are so imminent that ultimate ruin can be foreseen, the logical method of reclamation is in providing adequate systems of drainage to carry off the excess of water and the accumulated salts. This is expensive, but it is the only thing in this case to hasten the slow processes of nature, which are entirely inadequate in the presence of the present methods of irrigation and of culture. Underdrainage is expensive, but it has amply repaid for the investment in other localities where land is worth no more than in the Yellowstone Valley. Any land which is worth $50 per acre could well afford to be taxed for underdrainage if it is necessary, as in many places in the Yellowstone Valley, to save the investment from utter annihilation. It may be too soon yet to urge an extensive system of underdrainage in the valley, but some small systems should certainly be introduced, if necessary by cooperation, for an object lesson when it is considered necessary and timely to protect against trouble or to reclaim lands already abandoned. The owners will then see that it is feasible to protect their lands and to reclaim, through underdrainage, those that are abandoned.

It has been pointed out already that there are some crops which can stand much larger percentages of alkali than others. It is quite possible that other valuable crops can be found or can be bred which will stand large quantities of alkali, but it is unfortunate, indeed, for a locality like the Yellowstone Valley, which is originally free from alkali, to accept such conditions resulting from their injudicious methods of irrigation and try to find crops which will thrive upon lands which have been unnecessarily injured.

It must not be assumed, however, that a thorough system of underdrainage relieves one from exercising care and judgment in applying water to the land. There is less immediate danger of ruining the land, to be sure, but there are two things to be considered, namely, that an excessive use of water means just so much loss to irrigation and so much less land which can be brought under the ditch, and also that in the removal of these salts by the flow of the seepage waters out through the drainage system large quantities of really valuable plant food are likely to be removed from the soil. The very accumulation of these soluble salts is due to the arid conditions of the climate. The great fertility of the soils results from the accumulation of these salts, and if we introduce

PLATE XVI.

THE MAIN IRRIGATING CANAL: WATER TAKEN OUT OF THE YELLOWSTONE RIVER ABOUT 35 MILES ABOVE THE PLACE WHERE THIS ILLUSTRATION WAS TAKEN.

PLATE XVII.

FEEDING OF SHEEP AND RANGE CATTLE DURING THE WINTER SEASON.

37

artificial drainage, which will tax the resources of the soil, we may remove in the course of a generation, or even in less time than this, the accumulated results of the changes of vast geologic ages in the disintegration of rocks. By overirrigation and underdrainage we may remove in a few years the very conditions which contribute to the wealth of the country in the fertility of the soil.

In taking up new land in the Yellowstone Valley the heavy gumbo soils should be underdrained at the time the first irrigation waters are applied to the land. Even if the system of underdrainage is not complete at the start, a sufficient amount of it should be put in to answer the purpose at the beginning, and so arranged that it can be extended and more laterals put in as time goes on and the necessity of it becomes apparent. It is too late to wait until the damage has been done, for the accumulation of salts themselves acts on the heavy gumbo soils and makes them more impervious to water and harder subsequently to drain. Great care must be taken in the application of water. As little as possible should be applied at each time, so that there shall be as little waste as possible to go off as seepage water. The surface then should be thoroughly cultivated, unless otherwise protected from evaporation by alfalfa or other close-growing crops, so as to reduce the loss of water from the surface to a minimum and prevent thereby the accumulation of salts at the surface.

The rise in the level of water in wells must be looked upon with uneasiness and guarded against with great care.

The conditions in the Yellowstone Valley are particularly simple, and the danger from the rise of salts may be easily controlled. These investigations show the cause of the trouble, the actual conditions over a small section of the valley, and point out the logical methods of preventing trouble and of redeeming the land after the trouble has come. The locality is fortunate indeed in having no great excess of alkali in the soils previous to irrigation, as occurs over such large areas in adjoining States. The question involved is a simple problem, well within the control of the intelligent land owners of the valley.